Are Cryptozoological Animals-Real or Imaginary?

What are Cryptozoological animals? They are very rare animals which might even be imaginary or fantastical.

In this book we examine a group of these top cryptids which may or may not be real. The history and observations of these animals comprise each chapter and an evaluation of the evidence for their existence is provided at the end of each chapter.

Are all of these sightings fantasy or do these animals really exist? And how can they hide from most of us who have actively looked for them in the past?

We also provide available pictures, videos, and witness testimonies about these Cryptids.

You really should enjoy reading this book and hopefully you will have enough information to make your own decision.

Are Cryptozoological Animals-Real or Imaginary?

Are Cryptozoological Animals-Real or Imaginary?

Copyright Page

The book is copyrighted for 2019

Are Cryptozoological Animals-Real or Imaginary?

The Crazy and Out of the Box Series Book 4

By Martin K. Ettington

All Rights Reserved USA 2019

ISBN: 9781709794216

Printed in the United States of America

Are Cryptozoological Animals-Real or Imaginary?

Are Cryptozoological Animals-Real or Imaginary?

Other books by Martin K. Ettington

Spiritual and Metaphysics Books:
Prophecy: A History and How to Guide
God Like Powers and Abilities
Enlightenment for Newbies
Removing Illusions to Find True Happiness
Using the Scientific Method to Study the Paranormal
A Compendium of Metaphysics and How to Guides (Six books together in one volume)
Love from the Heart
The Enlightenment Experience
Learn Your Soul's Purpose
Pursuing Enlightenment
A Modern Man's Search for Truth
Use Intuition and Prophecy to Improve Your Life
The Handbook of Spiritual and Energy Healing

Longevity & Immortality:
Physical Immortality: A History and How to Guide
The Commentaries of Living Immortals
Records of Extremely Long Lived Persons
Enlightenment and Immortality
Longevity Improvements from Science
The 10 Principles of Personal Longevity
Telomeres & Longevity
The Diets and Lifestyles of the Worlds Oldest Peoples
The Longevity Six Books Bundle

Science Fiction:
Out of This Universe
Personal Freedom-Parts 1 & 2
The Psychic Soldier Series:
   Book 1-Himalayan Journey
   Book 2-A Soldier is Born
   Book 3-Fighting For Right
   Book 4-Earth Protector
The Immortality Sci Fi Bundle

The God Like Powers Series:
Human Invisibility
Invulnerability and Shielding
Teleportation
Psychokinesis
Our Energy Body, Auras, and Thoughtforms

The God Like Powers Series—Volume 1 Compilation

The Yoga Discovery Series:
Yoga-An Ancient Art Form
Hatha Yoga-Helping you Live Better
Raja Yoga-Through the Ages
The Yoga Discovery Package

Business & Coaching Books:
Creating, Publishing, & Marketing Practitioner Ebooks
Building a Successful Longevity Coaching Business
Why Become a Coach?
The Professional Coaching Success Trilogy
2020-Make Money Writing and Selling Books
The 2020 Handbook of High Paying Work Without a College Degree

Science, Technology, and Misc.
Future Predictions By and Engineer & Seer
The Unusual Science & Technology Bundle
The Real Atlantis-In the Eye of the Sahara
Are Cryptozoological Animals Real or Imaginary?
Real Time Travel Stories From a Psychic Engineer
Removing Limits On Our Consciousness-And Thinking Outside the Box
33 Incredible True Survival Stories
How to Survive Anything: From the Wilderness to Man Made Disasters
All About Mars Journeys and Settlement
Mining the Asteroid Belt

Ancient History
The Real Atlantis-In the Eye of the Sahara
Ancient & Prehistoric Civilizations
Ancient & Prehistoric Civilizations-Book Two
The History of Antediluvian Giants
The Antediluvian History of Earth
Ancient Underground Cities and Tunnels
Strange Objects Which Should Not Exist
Strange and Ancient Places in the USA
A Theory of Ancient Prehistory And Giant Aliens

Aliens and Space
Aliens and Secret Technology
Aliens Are Already Among Us
Designing and Building Space Colonies
Humanity and the Universe

Are Cryptozoological Animals-Real or Imaginary?

All About Moon Bases
All About Mars Journeys and Settlement
The Space and Aliens Six Books Bundle
A Theory of Ancient Prehistory and Giant Aliens

The Space Colonies and Space Structures Coloring Book
All About Asteroids

<u>The Longevity Training Series</u>

(A transcription of the online Multimedia Longevity Coaching Training Program)

The Personal Longevity Training Series-Book1-Long Lived Persons
The Personal Longevity Training Series-Book2-Your Soul's Purpose
The Personal Longevity Training Series-Book3-Enable Your Life Urge
The Personal Longevity Training Series-Book4-Your Spiritual Connection
The Personal Longevity Training Series-Book5-Having Love in Your Heart
The Personal Longevity Training Series-Book6-Energy Body Health
The Personal Longevity Training Series-Book7-The Science of Longevity
The Personal Longevity Training Series-Book8-Physical Body Health
The Personal Longevity Training Series-Book9-Avoiding Accidents
The Personal Longevity Training Series-Book10-Implementing These Principles

The Personal Longevity Training Series-Books One Thru Ten

These books are all available in digital and printed formats from my
website and on Amazon, Barnes & Noble, Apple ITunes, and many other sites

My Books Website is: http://mkettingtonbooks.com

Are Cryptozoological Animals-Real or Imaginary?

Signup for our Mailing List to get the following:

1) A discount coupon for 25% discount on all books on our site

2) Occasional Notices of new books available

3) Occasional Email on other offerings of ours (Monthly)

Go to this link to sign-up:

http://personal-longevity.com/mkebooks/emailsignup/

And click this link to get the FREE 102 page Ebook titled "Secrets of Many Things"

If you have any questions about this book or other subjects please contact the Author at:

mke@mkettingtonbooks.com

Are Cryptozoological Animals-Real or Imaginary?

Are Cryptozoological Animals-Real or Imaginary?

*Table of Contents

1.0 Introduction 1

2.0 Bigfoot/Sasquatch 3

3.0 Thunderbirds/ Pterodactyls 13

4.0 The Bunyip 25

5.0 Real Dragons 29

6.0 Really Giant Snakes 47

7.0 Mokele-Mbembe/Dinosaurs 53

8.0 Mongolian Death Worms 59

9.0 Unicorns 63

10.0 Werewolves 73

11.0 The Mapinguari 83

12.0 Sea Serpents 87

13.0 The Chupacabra 97

14.0 Carnivorous Trees and Plants 101

15.0 Summary 107

16.0 Bibliography 109

17.0 Index 113

Are Cryptozoological Animals-Real or Imaginary?

Are Cryptozoological Animals-Real or Imaginary?

1.0 Introduction

This is a book on Cryptozoology which means it is all about extremely rare animals which might exist or might even be imaginary.

This subject has always been fascinating to me. What if there are some dinosaurs living today in the deep jungles of Africa which science has never proven to exist? This statement sounds outlandish but we have found ancient fish in the sea who were supposed to have died out millions of years ago. A good example is the Coelacanth which was originally only known from fossils and was thought to have died when the dinosaurs died sixty some millions of years ago.

The coelacanth, which is related to lungfishes and tetrapods, was believed to have been extinct since the end of the Cretaceous period. More closely related to tetrapods than to the ray-finned fish, coelacanths were considered transitional species between fish and tetrapods. On 23 December 1938, the first Latimeria specimen was found off the east coast of South Africa, off the Chalumna River (now Tyolomnqa). Museum curator Marjorie Courtenay-Latimer discovered the fish among the catch of a local angler. Here is a picture of one:

Are Cryptozoological Animals-Real or Imaginary?

There are literally hundreds of animals which people have claimed to have seen so I'm only covering a few of the best known cases for which there is some evidence that they really exist.

Also, some people wonder if these strange and extremely rare animals have come to us from another dimension or traveled through time if they are real. We will assume for the purpose of this book that if these animals are real that they live in undiscovered nests and places but are real residents of this Earth.

I'm not including some animals such as the "Loch Ness Monster" which has had books and articles written about it for decades. We will mainly be covering those animals which might be proved to exist but which are lesser known.

One of my main goals here is to provide an open minded evaluation of the reality of the animals in question based on publically available information. At the end of each chapter there is an evaluation section where I draw my conclusions based on the available evidence.

I also don't' believe that many people just make these things up. Most of their reports are very emotional and they obviously saw something.

We need to be open minded and give the benefit of the doubt to these reports.

Are Cryptozoological Animals-Real or Imaginary?

2.0 Bigfoot/Sasquatch

Also known as the Yeti or Yowie depending on the part of the world you are in, this large hairy mammal has a huge number of reports and claims of encounters, but none have been accepted by scientists.

I decided to make Bigfoot the first Cryptid in this book because it also has the most sightings all over the world. Here is a map of sightings in North America alone:

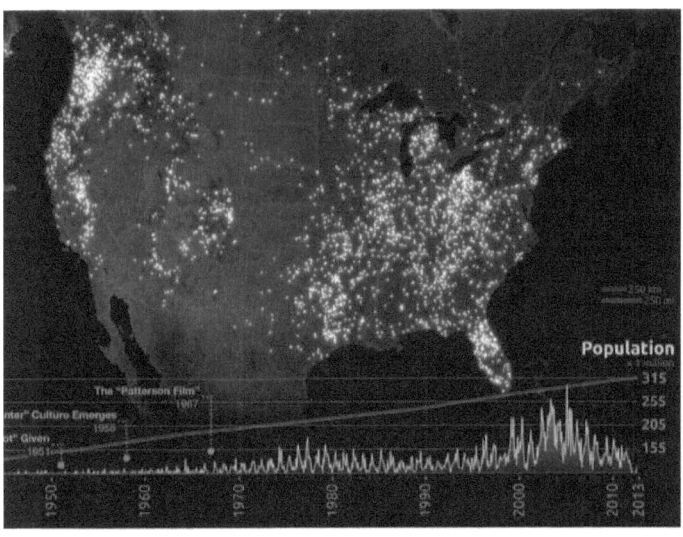

Reported sightings of Bigfoot — the legendary apelike creature that's been a favorite of cryptozoologists for decades — have abounded for decades. Now, for the first time, someone has created a map showing the places where alleged Bigfoot sightings have occurred.
Joshua Stevens, a doctoral candidate at Pennsylvania State University, used data compiled by the Bigfoot Field Researchers Organization (BFRO), which tries to

Are Cryptozoological Animals-Real or Imaginary?

document "the presence of an animal, probably a primate, that exists today in very low population densities," according to the group's website.

Stevens converted the BFRO data and, using geographic-information software, plotted 3,313 data points showing where people have claimed to see Bigfoot (aka Sasquatch, Skunk Ape, Yeti, Skookum or dozens of other names)

"Right away, you can see that sightings are not evenly distributed," Stevens said on his website. "There are distinct regions where sightings are incredibly common, despite a very sparse population. On the other hand, in some of the most densely populated areas, Sasquatch sightings are exceedingly rare. The terrain and habitat likely play a major role in the distribution of reports."

The map, which uses reports from 1921 to 2012, shows a plethora of supposed sightings in the Pacific Northwest, the Ohio River Valley, central Florida, the Sierra Nevada mountain range and the Mississippi River Valley.

Stevens' analysis also includes a chronological timeline showing a rise in reported sightings in the late 1970s (perhaps coinciding with the release of several B-movies about the mythical creature). Another spike in reported Bigfoot sightings occurred between 2000 and 2009.

Goodall, in an interview that was broadcast on NPR in 2006, said, "I'm sure that they exist." The famed primate researcher also confessed, "Well, I'm a romantic, so I always wanted that."

A handful of other academics, including Jeffrey Meldrum, professor of anatomy and anthropology at Idaho State

Are Cryptozoological Animals-Real or Imaginary?

University in Pocatello, have taken a scientific interest in the legend of Bigfoot, but to date, no hard evidence of any hominid or apelike creature has been substantiated.

All alleged samples of Bigfoot hair, for example, have turned out to be from elk, bears or cows. Photos, audio and film footage have been determined to be inconclusive or hoaxes, and no bodily remains have ever been found — despite the fact that there would have to be hundreds or thousands of the creatures in existence in order to maintain the "species."

Here are some of the best Bigfoot Sightings:

For decades, people around the world have been fascinated with the legend of Bigfoot, Sasquatch, yeti, or whatever you choose to call it. Sightings of a furry, upright biped and reports of beastly footprints have been reported from as far afield as the Himalayas. Although no definitive proof exists, the (often questionable) reports continue adding up. Outside plunged into the deep, dark corners of this subculture to compile the most famous—perhaps most convincing—Bigfoot photos ever captured. Here's the evidence. Is Bigfoot real? You be the judge.

Photo: Arguably the most famous and influential Bigfoot footage is the 1967 film shot by Roger Patterson and Bob Gimlin in Northern California. The "Bigfoot walk" it depicts has been parodied by many, but never truly replicated. Even with this enhanced image, it's hard to tell if it's a person in a gorilla suit or the real deal.

A youth group was camping in the Marble Mountain Wilderness when leader Jim Mills noticed a strange-looking creature skulking along a ridge nearby. He filmed it for nearly seven minutes, making the somewhat-grainy footage the longest video of an alleged Bigfoot sighting.

Are Cryptozoological Animals-Real or Imaginary?

British explorer Eric Earle Shipton snapped this photo while trekking through the Himalayas in 1951, alleging that the footprint belonged to a Yeti. In 2014, Christie's Auction house in London capitalized on the worldwide interest in Bigfoot and sold the original photo for nearly $5,000

Called "The Independence Day" film, this remarkably clear video shows an alleged adult Bigfoot walking through the woods, with a cub in tow. The filmmaker and exact location are unknown, and many skeptics claim that there is a telling visible seam of a gorilla suit.

In October 2012, a group of siblings hiking in Provo Canyon thought they spotted a bear in the woods and started filming. When the creature stood up on two legs, the hikers ran—abruptly ending the shaky video. A year later, the siblings launched a Kickstarter campaign to investigate other Utah Bigfoot sightings.

Are Cryptozoological Animals-Real or Imaginary?

In 2007, hunter Rick Jacobs captured some of the most famous Bigfoot images to date with a camera mounted to a tree in Pennsylvania's Allegheny National Forest. His camera also captured clear photos of bear cubs, offering evidence that the unidentified animal was not ursine. But skeptics believe the animal is just a bear sick with mange.

Locals in Johnstown, Pennsylvania, were baffled by a footprint measuring 17.75 inches found near a residential home in 1980. According to the Associated Press, the footprint coincided with reports of strange noises and a strong but unusual odor in the area.

Are Cryptozoological Animals-Real or Imaginary?

In 1994, former U.S. Forest patrolman Paul Freeman claimed he saw a family of Bigfoots in Washington's Blue Mountains. The video is shaky and grainy, but has been deemed the real deal by so-styled Bigfoot experts.

Mississippi resident Josh Highcliff captured video of this potential Bigfoot while hunting on his property in 2013. Afraid to go back to the woods, he posted the footage to

Are Cryptozoological Animals-Real or Imaginary?

YouTube asking for help to identify the animal or for a prankster to come forward.

A hiker was walking through the Utah Hills near Provo Canyon in 2012 when he spotted a large animal in the woods. As he approached, the animal stood up on two legs and started throwing rocks at him—supposedly a trademark behavior of sasquatches.

Evaluation of the Evidence:

Bigfoot has been sighted thousands of times in the United States, and all over the world. How could something seen so often still be considered fictional? Also, consider the number of pictures and videos which further lends credence to the existence of Bigfoot.

On the other hand, there are no confirmed remains of Bigfoot. I.E. no bones have been found. How could this be with all of these sightings?

Are Cryptozoological Animals-Real or Imaginary?

These two facts are real contradictions. One unusual possibility is that the Bigfoots have a unique way of hiding themselves at will such as a camouflage ability or psychic skill in hiding their visibilities. I know these are pretty far out ideas but they should be mentioned.

We don't really know the answers to all of these questions. Based on some of the dramatic sightings, videos, and pictures I have to come down on the side of those claiming to have seen Bigfoot and state my conclusion that Bigfoot is real.

Are Cryptozoological Animals-Real or Imaginary?

3.0 Thunderbirds/ Pterodactyls

Could there be modern day giant birds or even Pterodactyls who have survived from the days of the dinosaurs into the current day?

The American Indians had stories of Thunderbirds. Here is a Thunderbird Indian carving from Wisconsin:

Are Cryptozoological Animals-Real or Imaginary?

Here is a modern reproduction of the Thunderbird painting on rocks above the Mississippi River in Illinois:

The picture below is from the American civil war but has never been confirmed as real. However, this would be hard to fake in that era unless a huge fake dummy of the bird was made.

Are Cryptozoological Animals-Real or Imaginary?

Another picture of a totally different type of bird but which also meets the criteria for a Thunderbird is below:

A picture taken in 1945 which is also either an extremely deformed modern bird or something which is new to us:

Are Cryptozoological Animals-Real or Imaginary?

Some stories of Thunderbird sightings follow….

Thunderbird Tries to Abduct a Boy!

One of the most recent and highly-debated Thunderbird encounters occurred on July 25, 1977, in Lawndale, Illinois. On this evening, three boys were playing in a backyard when a pair of incredibly large birds swooped down on them. Two of the boys ran away unharmed, but they watched in horror as one of the birds picked up their friend, ten year old Marlon Lowe, in its claws and lifted the boy off the ground. The boy screamed and struggled, alerting his mother who was in the house at the time. She ran outside, yelling at the bird. After carrying the boy some distance away, it finally dropped him about two feet onto the ground. Lowe had deep scratches in his shoulder where the bird's talon dug into him.

The boy's mother reported the incident to the police, but her story was met with ridicule and disbelief. Although many experts reviewing the case believe that Marlon Lowe was attacked by a turkey vulture, others note the similarity in appearance between what the boys and the mother saw and an Andes condor, the much larger relative of the California condor.

Are Cryptozoological Animals-Real or Imaginary?

A Thunderbird Straight Out of Jurassic Park Reported in Alaska

Witnesses to a Thunderbird sighting in 2002 in Alaska claimed the bird they saw looked like "something out of Jurassic Park." According to these witnesses, the huge bird had a reptilian appearance and a 14-foot wingspan. It was larger and distinctive enough that it is unlikely that the witnesses mistook it for a seagull or eagle. The Anchorage Daily News even ran stories on the sightings.

Are Cryptozoological Animals-Real or Imaginary?

As sightings of giant birds pop up from time to time, it is hard not to draw connections between them and the Native American legends of Thunderbirds. The myths may turn out to be true.

Native American Thunderbird History:

The Thunderbird is a widespread figure in Native American mythology, particularly among Midwestern, Plains, and Northwest Coast tribes. Thunderbird is described as an enormous bird (according to many Northwestern tribes, large enough to carry a killer whale in its talons as an eagle carries a fish) who is responsible for the sound of thunder (and in some cases lightning as well.) Different Native American communities had different traditions regarding the Thunderbird. In some tribes, Thunderbirds are considered extremely sacred forces of nature, while in others, they are treated like powerful but otherwise ordinary members of the animal kingdom. In Gros Ventre tradition, it was Thunderbird (Bha'a) who gave the sacred pipe to the people. Some Plains tribes associated

Are Cryptozoological Animals-Real or Imaginary?

thunderbirds with the summer season (in Arapaho mythology, Thunderbird was the opposing force to White Owl, who represented winter.)

Thunderbirds are also used as clan animals in some Native American cultures. Tribes with Thunderbird Clans include the Kwakiutl and Ho-Chunk tribes. On the Northwest Coast, the thunderbird symbol is often used as a totem pole crest.

Some Indian stories of Thunderbirds:

(Potawatomi Legend)

Now regarding the Thunder Mountain in the western part of Marinette County: Thunder is a large bird like an Eagle, only much larger. And when this bird was created it was made to have power in order to defend us from the great serpents, who wanted to kill and eat the human race. It was also to moisten the earth for vegetation. Thunderers, we call these great birds. One of them is called Chequah. And the mountain we call Bikwaki, so Thunder Mountain is Chequah Bikwaki.

Many, many years have gone by since the Hill received its name. In the beginning of its Indian history the Thunderbirds used to make their nests here and sit on their two eggs until their young were hatched. Some Indians many years ago in the summer time visited the Hill and were surprised to find several pairs of young Thunders. It was always the custom with Indians to offer tobacco for friendship and safety.

And later on in another visit by the Indians a pond was discovered on the top of the Hill. And it was dangerous. The Serpent who lives under the Hill had caused this to be so that he could sun himself when the sky was clear. And on a sunny clear day he was sunning, probably asleep, when a lone Thunder discovered him and decided to catch him alive and carry him off. So the Thunder came down from the sky and caught the Serpent. The Thunder would carry him high. The Serpent, struggling, would carry the Thunder back down on the pond.

At that time an Indian hunter who was passing happened to look to the top of the Hill and to his surprise saw the two struggling, and went up to witness the great fight. He was noticed by them, and the Thunderbird spoke and said, "My friend, help me, and shoot the Serpent with your arrow, and I will make you a great man!" The Serpent also spoke and said, "Help me, and shoot the Thunder, and I'll promise you my friendship to the end of all time!" The Indian did not know which one to help, so he shut his eyes and shot an arrow toward the fighters and shot the Thunder. That shot weakened the Thunder and he fell down and was taken under the Hill as a prisoner. The Thunderbird is still there, and the Hill is called Chequah Bikwaki. Whenever there is going to be a thunderstorm lightning is seen flashing from the Thunder Mountain.

(Tales from the Hoh and Quileute)

HUNTERS FIND THE THUNDERBIRD

(Told by Frank Bennett. A Hoh myth A. B. R.)

Some men were hunting on the Hoh mountains. They found a hole in the side of the mountain. They said, "This is thunderbird's home. This is a supernatural place." Whenever they walked close to the hole they approached his place. He did not want any person to come near his house. He caused ice to come out of the door of his house.

Are Cryptozoological Animals-Real or Imaginary?

Whenever people came near there, he rolled ice down the mountain side while he made the thunder noise. The ice would roll until it came to the level place where the rocks are. There it broke into a million pieces, and rattled as it rolled farther down the valley. Everyone was afraid of Thunderbird and of the thunder noise. No one would sleep near that place over night.

Thunderbirds in different Indian Tribes:

Thunderbird of the Menominee Peoples

According to the Menominee tribe, the Thunderbirds live on an enormous mountain that floats in the sky. These majestic creatures are known to control the elements (rain, hail, etc.) and sometimes watch the happenings of human life. They are said to find great pleasure in fighting and the accomplishment of impressive feats. These Thunderbirds are known to be enemies of the Misikinubik (The Great Horned Snake) and are the reason mankind has not been devoured or overrun.

The Menominee Thunderbirds are also known to be messengers of the Great Sun and were highly respected by these peoples.

Thunderbird of the Ojibwe Peoples

The Ojibwe legends of the Thunderbird claim this creature was created by Nanabozho (one of the culture's hero figures) in order to protect people from evil underwater spirits. They lived in the four directions and migrated to the land of the Ojibwe during the spring with other birds. During this time they fought the underwater spirits. They stayed until the fall when the most dangerous season for the underwater spirits had passed. In the fall, they migrated south with other birds.

The Ojibwe Thunderbird legends also suggest that these creatures were responsible for punishing humans who broke moral rules. As the anger of the Thunderbird is known to be extreme, this would have been great incentive to maintain good moral conduct.

Thunderbird of the Winnebago Peoples

The Thunderbird of the Winnebago people suggests that this creature also had the power to grant people great abilities. Their traditions dictate that any man who has a vision of the Thunderbird during a fast will one day become a mighty war chief.

Thunderbird of the Sioux Peoples

Sioux legends claim that the Thunderbird was a noble creature that protected humans from Unktehila during the 'old times.' The Unktehila were said to be extremely dangerous reptilian monsters – without the help of the Thunderbird it is uncertain if man would have been able to overcome these creatures alone.

Thunderbird of the Arapaho Peoples

Arapaho mythology sees the Thunderbird as a summer creature (as did many of the tribes of the Great Plains). According to their legends, the Thunderbird was an opposing force to the White Owl (the creature that represented winter).

Are Cryptozoological Animals-Real or Imaginary?

Thunderbird of the Algonquian Peoples

The Algonquian Peoples had deep reverence for the Thunderbird in their culture. According to their legends, Thunderbirds were ancestors of the human race. Their stories often tell of the Thunderbird's part in the creation of the universe.

According to their myths, Thunderbird ruled over the upper world and the Great Horned Serpent ruled over the underworld. Thunderbird protects humans from the Great Horned Serpent and its followers by throwing lighting at underwater creatures.

Thunderbird of the Shawnee Peoples

Like many other tribes, the Shawnee people also believed the Thunderbird could change its appearance in order to interact with people. Their beliefs, however, detail that Thunderbirds appeared as boys and could be identified by their tendency to speak backwards.

Evaluation of the Evidence:

Similarly to Bigfoot sightings there have been many Thunderbird sightings over the years. The American Indians also have many legends of Thunderbirds as they do of the Sasquatch.

Also, we don't have any remains of dead Thunderbirds. We do have photos of two types of Thunderbirds—those that appear to be living Pterodactyls, and those which seem to be an extremely large modern regular bird with feathers.

Are Cryptozoological Animals-Real or Imaginary?

I'm willing to give the benefit of the doubt to the Thunderbird existing because of sightings over hundreds of years. People have seen some type of extremely large bird or flying dinosaur.

4.0 The Bunyip

During the early settlement of Australia by Europeans, the notion became commonly held that the bunyip was an unknown animal that awaited discovery. Unfamiliar with the sights and sounds of the island continent's peculiar fauna, early Europeans believed that the bunyip described to them was one more strange Australian animal and they sometimes attributed unfamiliar animal calls or cries to it. Scholars suggest also that 19th-century bunyip lore was reinforced by imported European folklore, such as that of the Irish Púca.

A large number of bunyip sightings occurred during the 1840s and 1850s, particularly in the southeastern colonies of Victoria, New South Wales and South Australia, as

Are Cryptozoological Animals-Real or Imaginary?

European settlers extended their reach. The following is not an exhaustive list of accounts:

Hume find of 1818

One of the earliest accounts relating to a large unknown freshwater animal was in 1818, when Hamilton Hume and James Meehan found some large bones at Lake Bathurst in New South Wales. They did not call the animal a bunyip, but described the remains indicating the creature as very much like a hippopotamus or manatee. The Philosophical Society of Australasia later offered to reimburse Hume for any costs incurred in recovering a specimen of the unknown animal, but for various reasons, Hume did not return to the lake. Ancient Diprotodon skeletons have sometimes been compared to the hippopotamus; they are a land animal, but have sometimes been found in a lake or water course.

Wellington Caves fossils, 1830

More significant was the discovery of fossilized bones of "some quadruped much larger than the ox or buffalo" in the Wellington Caves in mid-1830 by bushman George Ranken and later by Thomas Mitchell. Sydney's Reverend John Dunmore Lang announced the find as "convincing proof of the deluge", referring to Biblical accounts of the Flood. But British anatomist Sir Richard Owen identified the fossils as the gigantic marsupials Nototherium and Diprotodon. At the same time, some settlers observed that "all natives throughout these ... districts have a tradition (of) a very large animal having at one time existed in the large creeks and rivers and by many it is said that such animals now exist."

First written use of the word bunyip, 1845

Are Cryptozoological Animals-Real or Imaginary?

In July 1845, The Geelong Advertiser announced the discovery of fossils found near Geelong, under the headline "Wonderful Discovery of a new Animal". This was a continuation of a story on 'fossil remains' from the previous issue. The newspaper continued, "On the bone being shown to an intelligent black, he at once recognized it as belonging to the bunyip, which he declared he had seen. On being requested to make a drawing of it, he did so without hesitation." The account noted a story of an Aboriginal woman being killed by a bunyip and the "most direct evidence of all" – that of a man named Mumbowran "who showed several deep wounds on his breast made by the claws of the animal".

The account provided this description of the creature:

The Bunyip, then, is represented as uniting the characteristics of a bird and of an alligator. It has a head resembling an emu, with a long bill, at the extremity of which is a transverse projection on each side, with serrated edges like the bone of the stingray. Its body and legs partake of the nature of the alligator. The hind legs are remarkably thick and strong, and the fore legs are much longer, but still of great strength. The extremities are furnished with long claws, but the blacks say its usual method of killing its prey is by hugging it to death. When in the water it swims like a frog, and when on shore it walks on its hind legs with its head erect, in which position it measures twelve or thirteen feet in height.

Are Cryptozoological Animals-Real or Imaginary?

Evaluation of the Evidence:

The Bunyip cryptid has plenty of sightings but not any available bones to study. A problem with many of these cryptid claims. There is one skeleton found which might be the Bunyip or some prehistoric animal.

Again, being partial to what the native aborigines claim, I will side with them. This seems to be an undocumented but real cryptid.

5.0 Real Dragons

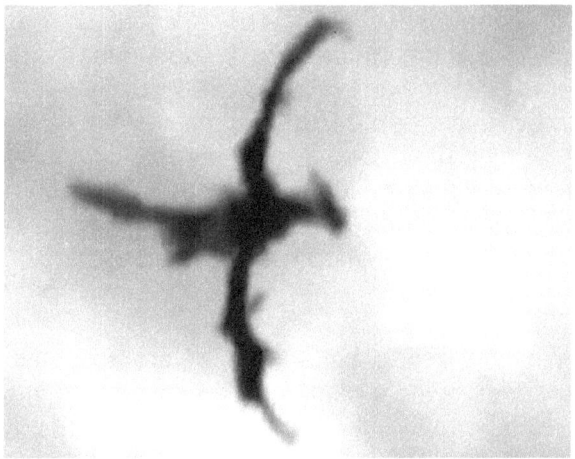

Are there real Dragons living on this Earth today? Most would think not since they are mostly creatures of legend. Although it is interesting that they exist both in Mesoamerican cultures and in China. Two places separated by many thousands of miles.

Reports of creatures very much like the fire-breathing, winged dragons of film and fantasy have been reported since far back in time, from civilizations all over the world.

One such very early account comes from England, and describes how the Briton king Morvidus was killed in 336 BC by a great dragon that rose from the Irish Sea and "gulped down the body of Morvidus as a big fish swallows a little one." The ancient explorer Titus Flavius Josephus also brought back tales of strange flying reptiles in ancient Egypt and Arabia, and the third century historian Gaius Solinus spoke of these creatures as well, further adding that they had potent venom that could kill a man even faster than he could realize that he had even been bit.

Many of the more spectacular early accounts of dragons were provided in the 4th century by Alexander the Great and his men after invading India. One account was reported by Alexander the Great himself, who claimed that he had seen an enormous hissing serpent lurking within a dank cave, and that the local tribes had worshiped it as a god, and his lieutenant, Onesicritus, also reported that there lived in India enormous serpents measuring 100 to 200 feet long.

This is very interesting, because there are accounts of such creatures in India going all the way back to the 1st century, when the Greek historian Strabo described fearsome winged reptiles in his book Geography: Book XV: On India, of which he says, "In India there are reptiles two cubits long with membranous wings like bats, and that they too fly by night, discharging drops of urine, or also of sweat, which putrefy the skin of anyone who is not on his guard."

Also from India is the account from the 3rd century historian Flavious Philostratus, who also claimed that India was home to dragons, and not only a habitat for them, but by his accounts absolutely crawling with them. He wrote in his The Life of Apollonius of Tyanna:

> *The whole of India is girt with dragons of enormous size; for not only the marshes are full of them, but the mountains as well, and there is not a single ridge without one. Now the marsh kind are sluggish in their habits and are thirty cubits long, and they have no crest standing up on their heads.*

Are Cryptozoological Animals-Real or Imaginary?

Some very intriguing early accounts of historical dragons can be found in the writings of the great 5th century B.C. Greek historian Herodotus, often referred to as "The Father of History" for his systematic method of recording events. According to the famous historian, these monsters lived in spice groves and frankincense trees, and he told that workers made a habit of driving them away with smoke before harvests, and Herodotus once wrote of these creatures:

> *There is a place in Arabia, situated very near the city of Buto, to which I went, on hearing of some winged serpents; and when I arrived there, I saw bones and spines of serpents, in such quantities as it would be impossible to describe. The form of the serpent is like that of the water-snake; but he has wings without feathers, and as like as possible to the wings of a bat.*

In the 8th century we have the curious account given by a St. John of Damascus, who wrote that during a battle against Carthage a huge dragon measuring 120 feet long had appeared behind the Roman army to approach them.

The army had then reportedly attacked and killed it, and had the skin sent to the Roman Senate, although what happened to it after that no one knows. This report is quite curious because it is a matter of fact account, without any obvious embellishment and sitting within other more mundane chronicles of the battle. He would even go as far as to state that these dragons were not magical creatures in any way, but rather just large, reptilian animals.

In later centuries we have the tales of the great explorer Marco Polo, who travelled around Asia, Persia, China, and Indonesia in the late 13th century and brought back all

Are Cryptozoological Animals-Real or Imaginary?

manner of fantastical tales of these exotic lands, their people, and their animals. Some of these reports included what can only be described as dragons. Within Polo's work The Travels of Marco Polo, there is a passage concerning a place in the Far East that he called "Karajan," which was apparently infested by the fierce beasts, and which he describes:

> *Here are found snakes and huge serpents, ten paces in length and ten spans in girth (meaning 50 ft. long and 100 inch circumference). At the fore part, near the head, they have two short legs, each with three claws, as well as eyes larger than a loaf and very glaring. The jaws are wide enough to swallow a man, the teeth are large and sharp, and their whole appearance is so formidable that neither man, nor any kind of animal can approach them without terror. Others are of smaller size, being eight, six, or five paces long.*

Again, this is all stated as fact, even going into depth about how the natives hunt and kill the creatures, and it is hard to know just what to make of it all. This apparently happens a lot with early dragon reports, and they even make appearances in respectable zoological compendiums. One good example of this can be seen within the pages of the work of Konrad Gesner, who was a great naturalist in the 16th century and wrote of dragons as if they were any other mundane animal, and gives one description of a beast seen in the 10th century of a dragon seen in Ireland with a horse-like head, a thick powerful tail, and stumpy, clawed legs.

Another famed 16th century naturalist by the name of Ulysses Aldrovandus also wrote seriously of dragons, and related several tales of the beasts, such as that of a herdsman who had been driving his herd of cattle in rural Bologna when he had encountered a small dragon that had blocked his path and hissed at him. The herdsman had then apparently killed the creature and saved the carcass.

Aldrovandus claimed to have come into possession of the body and to have even had it mounted, and spends a lot of time contemplating this specimen, speculating that it had been a juvenile dragon. Where the body went is anyone's guess, but Aldrovandus did have a watercolor portrait made of it. The 16th century is actually a treasure trove of real dragon encounters. In 1543 the historian Gesner wrote of a dragon-like creature in Germany, which he describes as having "feet like lizards, and wings after the fashion of a bat, with an incurable bite."

The historian and author Charles Gould would write of another historical case of the era concerning a man named Cardan, of which he says:

Are Cryptozoological Animals-Real or Imaginary?

Cardan states that when he resided in Paris he saw five winged dragons in the William Museum; these were biped, and possessed of wings so slender that it was hardly possible that they could fly with them. Cardan doubted their having been fabricated, since they had been sent in vessels at different times, and yet all presented the same remarkable form.

Bellonius states that he had seen whole carcases of winged dragons, carefully prepared, which he considered to be of the same kind as those which fly out of Arabia into Egypt; they were thick about the belly, had two feet, and two wings, whole like those of a bat, and a snake's tail.

Another rather interesting description of dragons was given in the early 16th century tome called the Aberdeen Bestiary, which goes into great depth on the appearance and behavior of the creatures and treats them as if they were all completely real. One passage reads:

> *The dragon has a crest, a small mouth, and narrow blow-holes through which it breathes and puts forth its tongue. Its strength lies not in its teeth but in its tail, and it kills with a blow rather than a bite. It is free from poison. They say that it does not need poison to kill things, because it kills anything around which it wraps its tail. From the dragon not even the elephant, with its huge size, is safe. For lurking on paths along which elephants are accustomed to pass, the dragon knots its tail around their legs and kills them by suffocation.*

Notice that it is explained rather matter-of-factly, with no attempt to really spruce it up with amazing imagery.

Are Cryptozoological Animals-Real or Imaginary?

Moving into the 17th century we have an account from 1619, in which a noble man named Christopher Schorerum saw a great flying dragon in Essex, England, of which he reported:

> *On a warm night in 1619, while contemplating the serenity of the heavens, I saw a shining dragon of great size in front of Mt. Pilatus, coming from the opposite side of the lake [or 'hollow'], a cave that is named Flue [Hogarth-near Lucerne] moving rapidly in an agitated way, seen flying across; It was of a large size, with a long tail, a long neck, a reptile's head, and ferocious gaping jaws. As it flew it was like iron struck in a forge when pressed together that scatters sparks. At first I thought it was a meteor from what I saw. But after I diligently observed it alone, I understood it was indeed a dragon from the motion of the limbs of the entire body.*

In 1658 there was published a book called Historie of Foure-Footed Beasts, which like some of the zoological compendiums we looked at earlier gave various descriptions of real animals and their behaviors. Once again, sitting there amongst the various detailed descriptions of known animals is a startlingly in-depth section on dragons, which explains them as it would any other normal animal. One passage reads:

> *This serpent (or dragon as some call it) is reputed to be nine feete, or rather more, in length, and shaped almost in the form of an axletree of a cart: a quantitie of thickness in the middest, and somewhat smaller at both endes. The former part, which he shootes forth as a necke, is supposed to be an elle [3 ft 9 ins or 1 I4 cms] long; with a white ring, as it*

were, of scales about it. The scales along his back seem to be blackish, and so much as is discovered under his belie, appeareth to be red… it is likewise discovered to have large feete, but the eye may there be deceived, for some suppose that serpents have no feete … [The dragon] rids away (as we call it) as fast as a man can run. His food [rabbits] is thought to be; for the most part, in a conie-warren, which he much frequents …There are likewise upon either side of him discovered two great bunches so big as a large foote-ball, and (as some thinke) will in time grow to wings, but God, I hope, will (to defend the poor people in the neighbourhood) that he shall be destroyed before he grows to fledge.

There be some dragons which have wings and no feet, some again have both feet and wings, and some neither feet nor wings, but are only distinguished from the common sort of Serpents by the comb growing upon their heads, and the beard under their cheeks. Gyllius, Pierius, and Gervinus . . . do affirm that a Dragon is of a black colour, the belly somewhat green, and very beautiful to behold, having a treble row of teeth in their mouths upon every jaw, and with most bright and clear-seeing eyes, which caused the Poets to say in their writings that these dragons are the watchful keepers of Treasures. They have also two dewlaps growing under their chin, and hanging down like a beard, which are of a red colour: their bodies are set all over with very sharp scales, and over their eyes stand certain flexible eyelids. When they gape wide with their mouth, and thrust forth their tongue, their teeth seem very much to resemble the teeth of wild Swine: And their necks

Are Cryptozoological Animals-Real or Imaginary?

> *have many times gross thick hair growing upon them, much like unto the bristles of a wild Boar. Their mouth, (especially of the most tamable Dragons) is but little, not much bigger than a pipe, through which they draw in their breath, for they wound not with their mouth, but with their tails, only beating with them when they are angry. But the Indian, Ethiopian, and Phrygian dragons have very wide mouths, through which they often swallow in whole fowls and beasts. Their tongue is cloven as it were double, and the Investigators of nature do say that they have fifteen teeth of a side. The males have combs on their heads, but the females have none, and they are likewise distinguished by their beards.*

It is all so painstakingly detailed and realistic one can clearly imagine exactly what they looked like. History is rife with accounts and reports such as these, and this has only scratched the surface of the countless such tales out there throughout the ages and from all over the world, stretching from Europe to the Middle East, Africa, and the Far East in places such as China, where dragons were a prominent feature of the landscape and revered.

Yet this is not a phenomenon merely confined to ages way back in the mists of time, not merely the constructs of simpler eras when people believed in myth, magic, and fairy tales, and dragons have continued to be reported up into more modern times.

Many of what is written of dragons in later years is not even all that spectacular or fantastical, such as the writings of Charles Gould, who documented many cases of dragons and spoke of them as being far from magical things of legend, but also very real. He would write in great detail on dragons in 1886, saying:

> *The dragon is nothing more than a serpent of enormous size; and they formerly distinguished three sorts of them in the Indies. Viz. such as were in the mountains, such as were bred in the caves or in the flat country, and such as were found in fens and marshes. The first is the largest of all, and are covered with scales as resplendent as polished gold. These have a kind of beard hanging from their lower jaw, their eyebrows large, and very exactly arched; their aspect the most frightful that can be imagined, and their cry loud and shrill... their crests of a bright yellow, and a protuberance on their heads of the colour of a burning coal. Those of the flat country differ from the former in nothing but in having their scales of a silver colour, and in their frequenting rivers, to which the former never come. Those that live in marshes and fens are of a dark colour, approaching to a black, move slowly, have no crest, or any rising upon their heads.*

Dragons have remained persistent right up to present day, and there are occasionally surprisingly recent sightings. In the early 1990s there was a report from a woman out hiking in the Rocky Mountains of Alberta and British Columbia, who says that she came across an actual dragon in the wilderness there, much to her disbelief. She says of her incredible experience:

> *The creature was in a beautiful shade of dark green and could easily blended with trees as he been standing by them but the witness report that he was perched on a rocky outcropping on the side of the mountain. He was fanning his wings slightly,*

Are Cryptozoological Animals-Real or Imaginary?

looking quite calmly into the valley below. I had been hiking up this mountain, when the movement of his head caught my eye. I had been this way before, and there was a group of trees on the cliff where there had been none before. I did not believe what I had seen at first, but the shape was too obvious, and he was parallel to me, about seven bus lengths away. I was climbing up one rock outface, he was on another. He was the most beautiful creature I had ever seen. His head was long, with a large eye ridge and two smaller bumps with a triceratops-like horn on his nose. At the back of his head were two large horns, jutting out backwards, and two smaller horns below them. They were a greyish-white and caught the light like dull silver. His forelegs were slightly smaller than his hind legs and were gripping the edge of the cliff. He looked as though he were a quadruped. He had slightly darker dorsal ridges running from between the longest horns to about halfway down his tail. As I stood there, gaping like a fish out of water, the dragon turned and looked at me. He cocked his head to the side, almost like a bird, then spread his enormous wings and vaulted off the cliff. He was absolutely elegant in the air, flapping his wings several times before banking into a glide and disappearing around the side of the mountain. My legs felt so weak that I had to sit down. I have been camping in those mountains for over ten years, and I had never seen anything to suggest that dragons might actually exist there. But after that encounter I began to think about it. What better place for a dragon to live than in the mountains? There are places in Banff and Jasper that nobody has ever been to, and there are many elk and deer and possibly even bears for it to feed on. Plenty of

Are Cryptozoological Animals-Real or Imaginary?

lakes, and the mountains themselves have many hidden caves and the like.

Even more recently, in 2001 an apparent dragon was purportedly seen by naturalists investigating a quarry in Wales. They described it was being "two and a half foot in length, serpentine dragon with four limbs and a head resembling that of a seahorse." The creature apparently hovered through the air without the aid of any noticeable wings, and the startled men watched it flit about for a full 4 minutes before it descended into one of the many dark caves dotting the area.

While it seems preposterous that the dragons we know from fiction, fairy tales, and fantasy could possibly have ever been real in any sense, the fact remains that remarkably similar stories have been reported throughout history by a wide range of disparate civilizations and cultures, so why is it that the dragon myths and tales are so universal? Could there have possibly have ever been anything to this? Theories have ranged from that these were just misidentifications and romanticized accounts of known animals, some form of outsized reptiles such as crocodiles or snakes, an undiscovered species, relic populations of dinosaurs surviving into modern times, perhaps even having evolved to their environment to take on a different appearance and abilities, or even as Carl Sagan once mused the constructs of some prehistoric shared racial memory infusing us.

Zoologist, Cryptozoologist, and researcher for the Center for Fortean Zoology, Richard Freeman, who has spent years studying historical accounts of real dragons for his book Dragons: More Than A Myth? has said of his own ideas on the matter:

Are Cryptozoological Animals-Real or Imaginary?

There are many creatures that have become linked to the lore and legend of what today we perceive and view as dragons, and some of these creatures are distinctly different to each other. But that should not take away from the fact that dragons are a real phenomenon. I am absolutely certain, having reviewed many ancient reports of dragon activity, that many sightings – perhaps two or three hundred years ago and probably further back – were genuine encounters, but where the witnesses were seeing what I believe to have been huge snakes, giant crocodiles, and something like the Australian 'monster lizard' Megalania.

In the end we have a phenomenon reported for over a millennium, of people of various cultures seeing these fierce reptilian beasts and it seems odd that they should all construct such similar legends and see such similar beasts in their respective histories. The dragon seems to be almost an archetype upon the landscape of the human psyche, somehow ingrained within us across cultures, and this makes it especially intriguing. Why should this be? Were dragons ever real in any sense, or are these just shared legends spewing forth from some universal sub consciousness? If they are real then what are they and do they exist now or have they gone extinct? With no real evidence and their tales doomed to mere speculation, it seems that we may never know the answers to these questions, and in the meantime the dragons must remain confined to legend, myth, and fiction.

The Feathered Serpent was a prominent supernatural entity or deity, found in many Mesoamerican religions. It was called Quetzalcoatl among the Aztecs, Kukulkan among the Yucatec Maya, and Q'uq'umatz and Tohil among the K'iche' Maya. Here is a carving of Quetzalcoatl in Mexico. It is certainly a dragon head:

Are Cryptozoological Animals-Real or Imaginary?

Also another Central American carving of the entire Dragon:

Are Cryptozoological Animals-Real or Imaginary?

Now we get to some frames from videos showing purportedly currently alive Dragons:

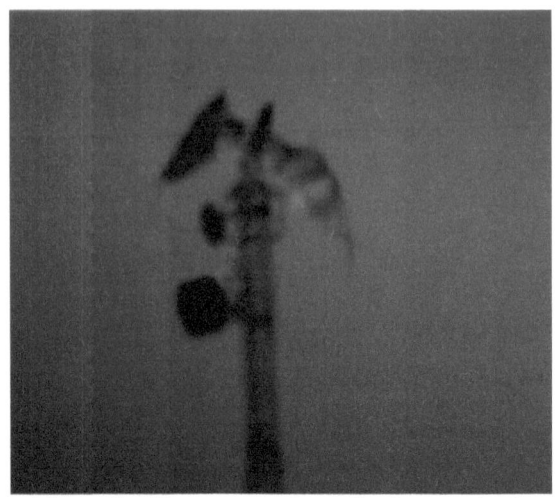

(From a video of a Dragon on top of a transmission tower)

(From another video showing this Dragon flying behind many rock spires)

Evaluation of the Evidence

Again we are in the situation where there are lots of sighting reports throughout history and even some pictures and videos of dragons from modern times.

Some writers have claimed to have seen dragon bones although those bones could have been misunderstood from really being Dinosaur bones.

In the end we have the same dilemma as with Bigfoot, and the Thunderbird. An animal reported throughout history for which we don't have a scientifically validated skeleton. So once more I will give the benefit to the numbers of observations and pictures to say YES—I believe in Dragons.

Are Cryptozoological Animals-Real or Imaginary?

6.0 Really Giant Snakes

The largest accepted record for the size of a snake is for a reticulated Python which was 30 feet long and weighed 350 pounds. But there are claims of snakes much larger...

In 1980 Van Lierde went on the British television show *Arthur C. Clarke's Mysterious World* to describe an encounter with a monstrous snake larger than any that have ever been observed.

The South American Annaconda is thought to be the largest snake in the World. During the show Van Lierde claimed that he saw the snake while flying over the Congo in a helicopter. He described the snake as about 50 feet long, with a triangular jaw. He said that he tried to approach the snake, but it raised up around ten feet, making a closer approach risky.

Are Cryptozoological Animals-Real or Imaginary?

Belgian fighter ace Remy Van Lierde in his RAF uniform with "Belgium" nationality title.

A passenger on the helicopter managed to take a picture of the snake, and although an exact measurement is impossible to make from just a photograph, it is clearly a remarkably large serpent.

Belgian pilot Remy Van Lierde claimed that while flying above the wilderness of the Katanga Province (in the southeast of the Belgian Congo) in 1959 that he took this photo of a giant snake.

However, there are some problems with this story. First off, the largest snake in the world, the Giant Anaconda, is only known to grow up to 30 feet. Second, such a large snake would have difficulty surviving due to its own weight. It would likely end up crushing its own organs. Third, even if such a large snake could support its weight, it certainly could not raise up ten feet in the air

Are Cryptozoological Animals-Real or Imaginary?

Giant Snake found on YouTube next to Amazon

Photograph shows 'giant snake' lurking in Borneo River
Villagers living along the Baleh river in Borneo fear a 100-foot snake could be lurking in the murky waters.

Are Cryptozoological Animals-Real or Imaginary?

Possible sighting of a giant snake in Borneo, Indonesia
12:43AM GMT 20 Feb 2009

An aerial photograph that appears to show a gigantic snake swimming along the remote waterway has emerged, sparking great concern among local communities.
But it is not clear whether the photograph is genuine, or a clever piece of photo-editing. Some suggested the 'snake' was in fact a log or a speed boat and others complained the colour of the river in the photo was too dark.

The most common theory is that the photo has been manipulated on a computer.

The image has even stumped the New Straits Times newspaper in Kuala Lumpur, which suggested readers decide for themselves.

However, on the banks of the river, villagers are convinced of the massive serpent's existence and have even given it a name, Nabau, after an ancient sea serpent which can transform itself into the shapes of different animals.

Evaluation of the Evidence

Snakes strike a visceral fear in most people. It might be since our small mammalian ancestors were common prey for snakes.

Also, many of them live in isolated jungle areas where people rarely travel. We know that snakes like to hide underground, underwater, and many come out only at night. Could there be snakes of much larger sizes like 50 to 100 feet long who have never been measured and legitimated?

Pictures of candidates for these huge snakes do exist. Due to their abilities to hide I have no problem believing that these huge snakes do really exist. Just don't get near enough to one to get eaten!

Are Cryptozoological Animals-Real or Imaginary?

7.0 Mokele-Mbembe/Dinosaurs

(Artist Rendering Above)

1909 saw the first mention of a brontosaurus-like creature in Beasts and Men, the autobiography of famed big-game hunter Carl Hagenbeck. He claimed to have heard from two independent sources about a creature living in Rhodesia which was described to them by natives as "half elephant, half dragon." Naturalist Joseph Menges had also told Hagenbeck about similar stories. Hagenbeck speculated that "it can only be some kind of dinosaur, seemingly akin to the brontosaurus." Another of Hagenbeck's sources, Hans Schomburgk, asserted that while at Lake Bangweulu, he noted a lack of hippopotami; his native guides informed him of a large hippo-killing creature that lived in Lake Bangweulu; however, as noted below, Schomburgk thought that native testimony was sometimes unreliable.

Are Cryptozoological Animals-Real or Imaginary?

Reports of entities described to be dinosaur-like in Africa caused a minor sensation in the mass media, and newspapers in Europe and North America carried many articles on the subject in 1910–1911; some took the reports at face value, others were more skeptical. According to German adventurer Lt. Paul Gratz's account from 1911:

> *The crocodile is found only in very isolated specimens in Lake Bangweulu, except in the mouths of the large rivers at the north. In the swamp lives the nsanga, much feared by the natives, a degenerate saurian which one might well confuse with the crocodile were it not that its skin has no scales and its toes are armed with claws. I did not succeed in shooting a nsanga, but on the island of Mbawala I came by some strips of its skin.*

Another report comes from German Captain Ludwig Freiherr von Stein zu Lausnitz [de], as described by Willy Ley in Exotic Zoology (1959). Von Stein was ordered to conduct a survey of German colonies in what is now Cameroon in 1913. He heard stories of an enormous reptile called "Mokéle-mbêmbe" alleged to live in the jungles, and included a description in his official report. According to Ley, "von Stein worded his report with utmost caution," knowing it might be seen as unbelievable. Nonetheless, von Stein thought the tales were credible: trusted native guides had related the tales to him, and the stories were related to him by independent sources, yet featured many of the same details. Though von Stein's report was never formally published, Ley quoted von Stein as writing:

> *The animal is said to be of a brownish-gray color with a smooth skin, its size is approximately that of*

an elephant; at least that of a hippopotamus. It is said to have a long and very flexible neck and only one tooth but a very long one; some say it is a horn. A few spoke about a long, muscular tail like that of an alligator. Canoes coming near it are said to be doomed; the animal is said to attack the vessels at once and to kill the crews but without eating the bodies. The creature is said to live in the caves that have been washed out by the river in the clay of its shores at sharp bends. It is said to climb the shores even at daytime in search of food; its diet is said to be entirely vegetable. This feature disagrees with a possible explanation as a myth. The preferred plant was shown to me, it is a kind of liana with large white blossoms, with a milky sap and applelike fruits. At the Ssombo River I was shown a path said to have been made by this animal in order to get at its food. The path was fresh and there were plants of the described type nearby. But since there were too many tracks of elephants, hippos, and other large mammals it was impossible to make out a particular spoor with any amount of certainty.

Are Cryptozoological Animals-Real or Imaginary?

Some additional eyewitness reports:

"I was in a boat on the river when I saw Mokele-mbembe. He began to chase us. Mokele-mbembe rose out of the water," one man told the BBC. "We ran, or he would have killed us."

Lake Tele, 5km across, is a hotspot for Mokele-mbembe sightings

Paul Ohlin, a community development worker who spent more than 10 years living with the Bayaka in Congo and the Central African Republic, just to the north, says the people who live in the area are in no doubt about the creature's existence.

"When people are sitting around the campfire talking, they talk about the Mokele-mbembe - it's something that's a reality in everyday life," he says.

At the same time he emphasizes their "spiritual connection" and "mystical relationship" with it.

The meaning of Mokele-mbembe

The standard translation for Mokele-mbembe is "one who stops the flow of rivers" - a reference to its purported penchant for nestling in the bends of rivers
But according to Paul Ohlin, it is also the word for "rainbow" and - crucially - for "mystery"
"The way they see the world is a little different to the way you and I see it," says Paul.

But their eyewitness reports still need to be taken seriously, in his view.

Are Cryptozoological Animals-Real or Imaginary?

"Certainly mythology surrounds it," says Adam Davies, a British man who spends his spare time and money travelling the world in search of undocumented species, and has twice gone to Africa on the trail of the Mokele-mbembe.

"But when you put it to people, 'Is this a real creature?' they become quite affronted… and they consistently came out with physical descriptions."

"Never dismiss tribal accounts on the basis that they must be talking tosh because they are tribal - that's not right and it's actually disrespectful," he says.

Evaluation of the Evidence

Lots of sighting claims for this dinosaur but no pictures. Native stories of encounters with this animal too.

I tend to believe the native claims. Many westerners poo-poo what natives claim but I think they are very reliable witnesses. I concur with their claims that this brontosaur is living in this remote area of Africa.

Are Cryptozoological Animals-Real or Imaginary?

8.0 Mongolian Death Worms

The worms are purportedly between two and five feet long (60 cm to 1.5 meters) and are thick-bodied.

In On the Trail of Ancient Man, Andrews cites Mongolian Prime Minister Damdinbazar who in 1922 described the worm:

> *It is shaped like a sausage about two feet long, has no head nor leg and it is so poisonous that merely to touch it means instant death. It lives in the most desolate parts of the Gobi Desert.*
>
> *In 1932, Andrews published this information again in the book The New Conquest of Central Asia, adding: "It is reported to live in the most arid, sandy regions of the western Gobi." Andrews, however, did not believe in the creature's existence.*

Are Cryptozoological Animals-Real or Imaginary?

Habitat and behavior

The worm is said to inhabit the western or southern Gobi. In the 1987 book Altajn Tsaadakh Govd, Ivan Mackerle described it as travelling underground, creating waves of sand on the surface which allow it to be detected. The Mongolians say it can kill at a distance, either by spraying a venom at its prey or by means of electric discharge. They say that the worm lives underground, hibernating most of the year except for June and July, when it becomes active. It is also reported that it most often comes to the surface when it rains and the ground is wet.

The Mongolians believe that touching any part of the worm will cause almost instant death and tremendous pain. It has been told that the worm frequently preyed on camels and laid eggs in its intestines, and eventually acquired the trait of its red-like skin. Its venom supposedly corrodes metal and local folklore tells of a predilection for the color yellow. The worm is also said to have a preference for local parasitic plants such as the goyo.

(A drawing of the worm in the above book)

Are Cryptozoological Animals-Real or Imaginary?

Czech cryptozoologist Ivan Mackerle is credited as being the foremost investigator of the death worm. He learned of the worm from a student and made the trip to southern Mongolia in 1990 to uncover more. His investigations were difficult, as he found many Mongolians were loath to speak of the legendary beast. Making it more complicated was an order by the Mongolian government outlawing searches for the death worm. Eventually the ban fell and Mackerle was able to search for answers.

In his book "Mongolské záhady" (Mongolian Mystery), Mackerle chronicled the worm from second-hand reports. The creature is described as a:

"sausage-like worm over half a meter (20 inches) long, and thick as a man's arm, resembling the intestine of cattle. Its skin serves as an exoskeleton, molting whenever hurt. Its tail is short, as if it were cut off, but not tapered. It is difficult to tell its head from its tail because it has no visible eyes, nostrils or mouth."

He never witnessed it himself, but Ivan Mackerle eventually determined the Olgoi-Khorkhoi could be real.

Evaluation of the Evidence

This is one case where I don't think we have enough information to make a call on the worm's reality or not.

Are Cryptozoological Animals-Real or Imaginary?

9.0 Unicorns

Are Unicorns Real? Most people would say NO they are just a childish fantasy.

Sometimes truth is stranger than fiction. The saola is a species that could be mistaken for a myth, but it's real.

Sometimes known as the "Asian unicorn," the saola is a creature related to wild cattle that lives in remote areas of Vietnam and Laos. It is seldom seen and was only discovered in 1992. The evidence then? A single skull with a pair of long, straight horns, according to the World Wildlife Fund. In 1999, a couple of saolas were seen on camera traps, proving that they still survived in the wild.

The next confirmed sighting wasn't until 2010, when villagers in Laos captured an adult male and brought it to an enclosure in their village. Sadly, the ordeal of captivity killed the saola, but the incident did prove that the animal still survived in the region. There are no saolas in captivity anywhere in the world, and the species is considered critically endangered.

There is perhaps no other fairy tale creature of lore as well-known and also beloved as the unicorn. Often depicted as a majestic white horse with a flowing mane and a single horn protruding from the center of its head, but also variously as a one-horned deer, ass, or goat, the unicorn has actually been depicted in one form or another for millennia and throughout cultures around the world. Such one-horned creatures have been depicted in art and legend since at least the Mesopotamian era (5000-3500 B.C.), and can be found in the legends of such far-flung countries as Greece, India, China, and several countries of Europe, especially in Scotland, where it is to this day designated as their national animal. Unicorns are mentioned in many sacred books around the world, including the Bible, where the creatures are mentioned numerous times. Unicorns in various cultures are typically said to have magical powers to some extent, and they were thus highly prized for these qualities. Other traditions regard them as signs of good luck, or conversely harbingers of tragedy.

Are Cryptozoological Animals-Real or Imaginary?

Considering that they are such a part of myth and legend, unicorns are considered by many to merely just that, and that they cannot possibly exist for real outside of the confines of folklore. Indeed the very word "unicorn" has become virtually synonymous with something which we want to exist, but which is either a myth or so rare to the point of being nearly nonexistent. Yet there have been numerous reports throughout history that suggest that they may have been perhaps real.

For instance, from around the 15th century in Europe there were often said to be "unicorn horns" brought forward, which were considered to be valued well beyond their weight in gold, and powdered unicorn horn was a popular, sought after item all the way up until the 18th century. The horn was said to have all sorts of beneficial effects, and could supposedly heal or cure all manner of injuries or ailments, as well as serve as an antidote for poison. Pope John III and King James I of England both allegedly paid great sums of money for whole unicorn horns, although it is not known what happened to them and these may have very well been the horns of the one-horned narwhales. Similarly, unicorn milk, tears and blood were also thought to be very real, and to have remarkable healing properties and even potent aphrodisiac qualities, which fetched them exorbitant prices.

Often even whole skeletons or other parts of unicorns would turn up, although these are mostly considered to have been cleverly crafted fakes, such as such a skeleton put together by German scientist Otto Von Guericke, who even provided a sketch of the creature in an otherwise serious book on natural history in the 1600s.

Sightings of unicorns go back even further, with one of the earliest seemingly realistic accounts coming from the Greek doctor Ctesias in the 4th century B.C., who wrote quite seriously of such creatures during his travels through Persia. In this case the animal was described as being a type of wild ass as large as a horse, which had a white body, red head, and bright blue eyes, with a multi-colored single horn sprouting from the head measuring around one and a half feet in length. These animals were described as being very powerful and fast, to the point that they were able to easily outpace any horse. There is nothing in the account to suggest that this was seen as anything other than a real, unidentified beast. The Greek historian Herodotus also wrote about the "horned

Are Cryptozoological Animals-Real or Imaginary?

ass" of Africa in the 3rd century B.C. Other travelers such as Pliny the Elder, as well as Marco Polo, who called them "ugly brutes," would also write of encountering unicorns during their journeys. Other famous historical figures who claimed to have encountered unicorns were Julius Caesar, Genghis Khan, and the Chinese philosopher Confucius, among others.

It has since been speculated that these early reports were most likely misidentifications of some other exotic one-horned animal such as the rhinoceros, or that they were misrepresentations of other horned animals such as oxen, bulls, ibexes, onyxes, or goats, perhaps stemming from exaggeration or sightings of deformed individuals with some form of genetic mutation, all of which could have been colored by local myth to fit in with unicorns. A mutation is perfectly feasible, as there have indeed been found one-horned specimens of these animals on occasion. Yet many of these witnesses would have been familiar with some of these other animals to some extent, and don't seem likely to have mixed them up with a supposedly mythical beast like the unicorn. What did they really see? Knowone really knows for sure.

Sporadic sightings of supposed unicorns have have continued right up into the modern day. In 1991, the renowned Austrian naturalist Antal Festetics allegedly saw a unicorn while filming a wildlife documentary in the Harz Mountains. He claimed in an interview with *Die Ganze Woch* magazine that he had been out on horseback one evening when he saw something he would never forget. He would say:

> *Suddenly a unicorn came towards me at a gallop. There was a glow of light around the animal. My horse reared and almost threw me. Then, just as quickly, it was gone.*

Are Cryptozoological Animals-Real or Imaginary?

He would tell this account more than once, and Festetics even said that he had videotaped the creature, although it is unknown where this footage got off to, and it never did make it into his documentary that he had been filming at the time. Speaking of photographic evidence, there have been several supposed photos and videos of unicorn sightings that have come out over the years. In 1968, a Robert Vavra claimed:

> *My first face-to-face encounter with a unicorn took place in a Mexican jungle near Tamazunchale in the spring of 1968. The one photograph that I was able to snap is reproduced on this page. This picture, as the reader can see, is so blurry and nondescript that had I attempted to publish it at the time, I would have been considered as much a*

Are Cryptozoological Animals-Real or Imaginary?

> *crank or hoaxter as those persons who occasionally submit to the press out – of – focus, suspect.*

A video of a supposed unicorn was taken in 2010 in the wetlands of the Don Valley, near Toronto, Ontario, Canada by a local birdwatcher named Peter Hickey-Jones. The video seems to show a white horse with a single horn emerge from the trees gallop though a creek and disappear into the woods again, and this footage was then purportedly submitted to the Ontario Science Center for analysis. When this video began making the rounds on the Internet it created quite a stir, and not surprisingly many immediately claimed it to be a hoax, especially since the video conveniently coincides with a promotional push for an exhibit at the center entitled "Mythic Creatures: Dragons, Unicorns and Mermaids."

Are Cryptozoological Animals-Real or Imaginary?

Even more recently is a sighting made in the highlands of Wick in Caithness, Scotland, in 2014. The unidentified witness claims that he was out hiking near the Castle of Old Wick when he saw what he described as "basically a horse with a horn," which wandered off towards Loch Hempriggs, to the south of the town. At the time the sighting became big news in the area, even spawning plans to launch a "unicorn safari" for any curiosity seekers that came in to investigate. It is unclear just what the witness saw or what became of this report, but it is certainly odd.

It seems that in this modern world there should be no place for such mythical creatures anymore, and that they cannot possibly be real in any sense. It seems absurd that anyone could possibly see a real-life unicorn, but there these reports are. One wonders if there is any grain of truth that can account for such reports throughout the ages. Were these misidentifications, mutations, or hoaxes? If they are real, what could they be? Are they an undiscovered species or subspecies? Are these unicorns perhaps something more magical in nature, or maybe something from another dimension seeping into our own? It is unknown, but certainly interesting to think about.

Evaluation of the Evidence:

The Unicorn is a magical animal if we believe the historical accounts of it. There are also a few sightings and even some pictures although the majority of what we know about Unicorns is obviously mythological.

Are Cryptozoological Animals-Real or Imaginary?

I think the Unicorn is some type of spiritual being which some people are able to see.

Is it a real animal on Earth? I wonder. Am open minded about this one but haven't formed a solid opinion yet.

Are Cryptozoological Animals-Real or Imaginary?

10.0 Werewolves

Werewolves are an ancient mythical animal which people have reported many scary tales regarding. One wonders if these tales might be a result of our encounters with wolves. Wolves rarely hurt but occasionally do attack humans. Also the fact that we have domesticated dogs for tens of thousands of years might be involved since subconsciously we might think that dogs could attack us. Here are some stories of werewolves…

<u>The Beast of Gévaudan</u>

(17th century woodcut depicting a hunt for the Beast of Gévauda.)

In the 18th century, the former French province of Gévaudan was terrorized by the so-called <u>La Bête du Gévaudan</u> (The Beast of Gévaudan). The Beast was first spotted by a woman tending cattle in the forest near Langogne in June. Her bulls scared it off, but not long after it attacked and killed a 14-year-old girl. Over the ensuing months, sightings and attacks mounted.

Those who had seen the Beast described a large wolf with unusual red fur streaked with black. And it was prolific. According to a 1980 study, there were 210 attacks in all, 113 of which were fatal.

In 1765, King Louis XV decreed that the French state would help slay the beast.

When the appointed professional wolf hunters, Jean Charles Marc Antoine Vaumesle d'Enneval and his son Jean-François failed to kill the Beast, the king sent Lieutenant of the Hunt François Antoine instead. Antoine slayed three giant grey wolves, yet the attacks still continued.

It wasn't until a local hunter named Jean Chastel shot a wolf on June 19, 1767 that the attacks were declared over. Nowadays, it is thought that the Beast of Gévaudan wasn't a single wolf at all, but many individual wolves. When France went on a wolf-killing rampage, these wolves were slain, one by one, until none were left and the attacks abated. Not that killer wolves were unusual. According to historian Jean-Marc Moriceau, some 7,600 people were killed by wolves in France between 1362 and 1918.

Are Cryptozoological Animals-Real or Imaginary?

The Livonian Werewolf

(Werewolf devouring children, Lucas Cranach the Elder, 1512.)

Werewolf confessions could be quite peculiar. Take Thiess of Kaltenbrun. Living in Swedish Livonia in the 17th century, Thiess was widely believed among his neighbours to be a werewolf who had dealings with the Devil.

Local authorities didn't much care. After all, Thiess was in his eighties. What harm could he do with a few tall tales? But when they brought him in for questioning on an unrelated matter in 1691, he voluntarily began divulging details of his werewolf lifestyle... although with many inconsistencies.

Are Cryptozoological Animals-Real or Imaginary?

According to his account, Thiess had given up lycanthropy 10 years prior to his appearance before the judges in 1691. Before that, he and other werewolves would change into wolves on St Lucia's Day, Pentecost and Midsummer Night by donning magical wolf pelts (although he later changed his story and said they just stripped naked and turned into wolves).

They would then maraud the countryside, killing farm animals and cooking and eating them (when asked how wolves cooked meat, he declared they were still human, not wolves).

His story only grew stranger. He claimed that werewolves were the agents of God, and would travel to hell to battle the Devil and his witches, bringing back grain and livestock the witches had stolen. In fact, he said, he had done so just one year earlier, contradicting his earlier claim of having renounced lycanthropy.

When it was revealed that Thiess was not a devout Lutheran, and indeed practiced a form of folk magic involving charms and blessings, the judges ordered Thiess flogged and exiled. What happened to the strange chap after that is unknown.

Are Cryptozoological Animals-Real or Imaginary?

The Wolf of Ansbach

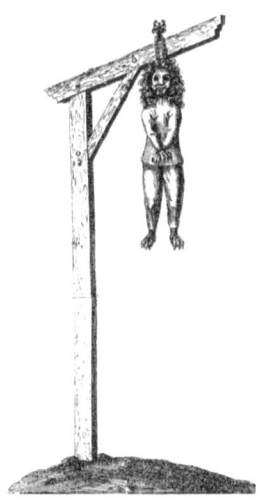

A wolf in magistrate's clothing: 1685 woodcut of the wolf on display.

In 1685, a wolf was terrorizing and killing humans in the town of Neuses in the Principality of Ansbach in what is now Germany. This was not unusual, but the town's chief magistrate Michale Leicht, had just died. He was a cruel and unpopular man, and it was said that the wolf visited Leicht's residence, so it was only a small leap for people to claim the wolf was Leicht, returned as a werewolf for his sins.

The wolf's death was not terribly eventful. The people organized a hunt and chased the wolf into a well and killed it. What they did with its body is pretty macabre, though. They paraded it through the streets, then prepared it for display. They cut off its muzzle, dressed it in human clothes and placed a wig on its head and a mask on its face, so that it resembled Leicht. They then hung the body from a gibbet so that everyone might enjoy the sight. After, some time, the wolf was removed from the gibbet, and its corpse preserved and put on permanent display at a local museum. Because that's not weird or creepy at all.

The Werewolf of Allariz

Widely thought of as Spain's first ever serial killer, Manuel Blanco Romasanta is unusual for a werewolf, operating late in the mid-19th century.

Actually, Romasanta was an unusual case in a few ways. Born in 1809, he had been raised as a girl until about the age of six, at which point doctors discovered he was male. He grew up, married and worked as a tailor. When his wife died in 1833, he took up the travelling salesman trade, also guiding travelers around Spain and Portugal.

His first known murder was Vicente Fernández, the constable of León. Fernández was found dead in 1844 after attempting to collect a debt from Romasanta. Rather than face the law, Romasanta fled to Portugal.

Are Cryptozoological Animals-Real or Imaginary?

During this time, he murdered several people who had hired him as a guide. He was not a cunning man. Romasanta was noticed selling their clothes, and rumors started to circulate that he was selling soap made with human fat. A complaint was lodged and Romasanta was arrested.

He confessed to 13 murders, but here is where it gets wolfish. He said he had been cursed with lycanthropy. But upon being asked to demonstrate his transformation abilities, Romasanta declared that the curse had passed and he was no longer afflicted.

He was actually acquitted of four of the deaths. Those, forensic examination found, had been committed by real wolves. However, he was found guilty of the rest. A phrenological examination of Romasanta by doctors determined that he had invented his "curse", and he was sentenced to death. This was commuted to life imprisonment on the request of a French hypnotist, who believed that Romasanta was suffering a delusion and petitioned a stay of execution so that he might study the man.

An 1863 newspaper reported that Romasanta passed away that year in prison from stomach cancer.

Are Cryptozoological Animals-Real or Imaginary?

The Werewolf of Bedburg

Composite woodcut by artist Lukas Mayer depicting the events of Stumpp's torture and execution.

One of the most famous werewolf cases is Peter Stumpp, a wealthy farmer accused of being a serial murderer, cannibal and werewolf in Rhineland in 1589.

In the years preceding Stumpp's arrest, the country town of Bedburg had been plagued with horrors. It started with dead and mutilated cattle, but bodies of townsfolk were also soon found in the fields. Initially, it was thought that a wolf or wolves were attacking, but the creatures evaded capture. Finally, in 1589, a hunting party managed to corner the wolf with its hounds. When the humans approached, they saw, according to reports, not a wolf at all. Instead, the hounds had cornered Stumpp.

The most damning piece of evidence was that Stumpp's left hand had been lopped off. The wolf had had its left forepaw cut off. Since wolf and man had the same injury, wolf and man must be one and the same.

Stumpp confessed, but it's a questionable confession at best. He had been subjected to torture, including the rack. He said he'd made a pact with the devil when he was 12. He had been given a magic belt which allowed him to turn into a wolf. He confessed to killing 14 children and 2 pregnant women. He ate of their flesh and ravished their bodies. He killed his own son, and had a sexual relationship with his own daughter.

He was sentenced to die in the most awful manner. He was fixed to a breaking wheel, and had flesh torn from his body with red-hot pincers. His limbs were broken with the blunt side of an axe so he might not rise from the grave. Finally, he was beheaded. His head was placed on a pole with the figures of a breaking wheel and a wolf on it, as a warning to others.

His daughter and mistress were also flayed, strangled and burned.

It is not known whether the crimes were truly committed by Stumpp. At the time, the region was deeply affected by the Cologne War. Stumpp was a Protestant convert, and the region had been seized by the Catholics in 1857. His death was to the Catholics' advantage, as his considerable wealth would fall to them. In addition, Stumpp's death could have served as a strong warning to other Protestants.

Are Cryptozoological Animals-Real or Imaginary?

Evaluation of the Evidence:

Wolves have been around with humans tens of thousands of years. Dogs have been shown genetically to be evolved from Wolves which can even breed with Dogs which shows how close they are genetically.

Although it is rare, Wolves have been known to attack and sometimes kill humans. So there is a basic fear there of wolves and of the idea that dogs could turn against humans because they are related to wolves.

However, there is no good evidence for humans shapeshifting into wolves and visa versa. Knowing about the paranormal world I do believe that shapeshifters may exist even though they are very rare.

Though this does not mean that there is a race of werewolves out there. I just do not believe in a race of werewolves even though there might be shapeshifter humans who have occasionally turned into wolves.

11.0 The Mapinguari

Zoologists generally regard the creature as fictional. Some biologists have speculated that the creature's place in Amazon folklore may stem from cultural memory of the now extinct giant ground sloth. Some researchers also have suggested that it sounds more like a bear than anything else. In 2007, American anthropologist and ethnobiologist Glenn Shepard Jr said, "At the very least, what we have here is an ancient remembrance of a giant sloth, like those found in Chile recently, that humans have come into contact with ... Let me put it this way: Just because we know that mermaids and sirens are myths doesn't mean that manatees don't exist."

American ornithologist David Oren proposes that the creature may represent a living ground sloth. As of 2001, Oren had spoken to between fifty and eighty indigenous Brazilians, rubber planters, and miners who claimed to have seen, and seven hunters who claimed to have shot, mapinguaris or other unknown animals with which Oren equates the mapinguari, in or near Eirunepé, Manicoré, and Carauarí in Amazonas; Marabá in Pará; the Parque Nacional da Serra do Divisor in Acre; Juína in Mato Grosso; and Tocantins. The seven hunters all claimed not to have saved any remains due to their terrible odour, which they claimed made them nauseous and light-headed.

According to Oren, he was told in the 1990s by the hunters that the mapinguari is an animal with long reddish, blackish, or brownish fur; a powerful build and a height of around 2 metres when standing on its hind legs; claws shaped like those of the giant anteater, but the size of a giant armadillo's; a muzzle like that of a horse, but shorter; a short but stout tail; four large canine teeth; a very strong and unpleasant smell compared to a mixture of faeces and rotting flesh; and the ability to walk both quadrupedally and bipedally. Six of the seven hunters claimed that they had to shoot the mapinguaris they killed in the head with special slugs of solid lead, whilst the seventh hunter claimed to have emptied his .38 caliber revolver into the animal's chest.

Oren initially thought the mapinguari must be a mylodontid ground sloth, suggesting that a mylodont's osteoderms, curved feet, and heavy tail tip could explain the mapinguari's more inexplicable purported characteristics of bulletproof skin, backwards feet, and bottle-shaped footprints.However, after receiving more details from the seven hunters who claimed to have shot specimens, Oren changed his view and theorised that the mapinguari would be a megalonychid, not a mylodontid, ground sloth, on account of its alleged canine teeth and locomotion.

Are Cryptozoological Animals-Real or Imaginary?

Evaluation of the Evidence:

The Mapinquari is an interesting case of an animal purported to be found in the wilds of South America. It is talked about in similar ways to the Dinosaur of Africa in that there are limited people who have seen it and it is also very similar to the ancient giant Sloth which lived within the life of mankind.

For this reason I think it probably exists as a rare creature in the back jungles of Brazil.

Are Cryptozoological Animals-Real or Imaginary?

Are Cryptozoological Animals-Real or Imaginary?

12.0 Sea Serpents

Sea Serpents have been reported of several types. Here are some of the claims…

The Kraken:

After returning from Greenland, the anonymous author of the Old Norwegian natural history work Konungs skuggsjá (circa 1250 A.D.) described in detail the physical characteristics and feeding behavior of these beasts. The narrator proposed there must be only two in existence, stemming from the observation that the beasts have always been sighted in the same parts of the Greenland Sea, and that each seemed incapable of reproduction, as there was no increase in their numbers.

> *There is a fish that is still unmentioned, which it is scarcely advisable to speak about on account of its size, because it will seem to most people incredible. There are only a very few who can speak upon it clearly, because it is seldom near land nor appears where it may be seen by fishermen, and I suppose there are not many of this sort of fish in the sea. Most often in our tongue we call it hafgufa ("kraken" in e.g. Laurence M. Larson's translation). Nor can I conclusively speak about its length in ells, because the times he has shown before men, he has appeared more like land than like a fish. Neither have I heard that one had been caught or found dead; and it seems to me as though there must be no more than two in the oceans, and I deem that each is unable to reproduce itself, for I believe that they are always the same ones. Then too, neither would it do for*

other fish if the hafgufa were of such a number as other whales, on account of their vastness, and how much subsistence that they need. It is said to be the nature of these fish that when one shall desire to eat, then it stretches up its neck with a great belching, and following this belching comes forth much food, so that all kinds of fish that are near to hand will come to present location, then will gather together, both small and large, believing they shall obtain their food and good eating; but this great fish lets its mouth stand open the while, and the gap is no less wide than that of a great sound or bight, And nor the fish avoid running together there in their great numbers. But as soon as its stomach and mouth is full, then it locks together its jaws and has the fish all caught and enclosed, that before greedily came there looking for food.

<u>Giant Squids:</u>

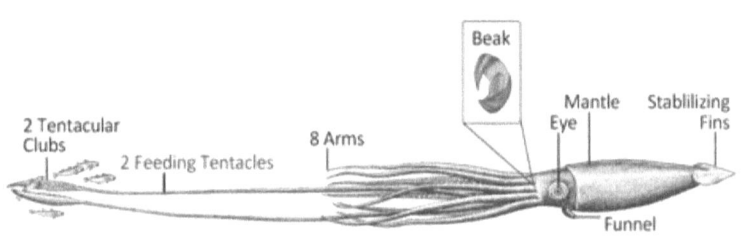

Are Cryptozoological Animals-Real or Imaginary?

What about the legends of Sea Serpents. Sailors have been reporting these for thousands of years. Is there some truth to these scary claims that a Sea Serpent threatened or even sunk a ship?

Giant squid live up to their name: the largest giant squid ever recorded by scientists was almost 43 feet (13 meters) long, and may have weighed nearly a ton. You'd think such a huge animal wouldn't be hard to miss. But because the ocean is vast and giant squid live deep underwater, they remain elusive and are rarely seen: most of what we know comes from dead carcasses that floated to the surface and were found by fishermen.

But after years of searching, in 2012 a group of scientists from Japan's National Science Museum along with colleagues from Japanese public broadcaster NHK and the Discovery Channel filmed a giant squid in its natural habitat for the first time. The species was first recorded live in 2006, after researchers suspended bait beneath a research vessel off the Ogasawara Islands to try and hook a giant squid. As the camera whirred, the research team pulled a 24-foot (7-meter) squid to the surface alive enabling people around the world to finally see a living, breathing giant squid.

Are Cryptozoological Animals-Real or Imaginary?

Other reports include:

In 1779 Edward Preble, an ensign on the ship Protector, spotted a great serpent in Penobscot Bay, off the coast of Maine. Ordered into a longboat, Preble approached the beast—whose head rose several feet out of the water atop a thick neck like that of a giant snake—took aim, and fired a shot. He missed. The creature vanished beneath the surface, never to be seen by Preble again.

Nearly forty years later, in the summer of 1818, a serpentine monster was sighted from Week's Wharf in Portland Harbor. In 1836 Captain Black, aboard the schooner Fox, reported a sighting of a snakelike creature in the seas near Mount Desert Rock. Major General H. C. Merriam and his sons were sailing to Wood Island Light in 1905, when they spotted a "monster serpent," which proceeded to swim circles around their boat. In 1910, from the deck of the steamer Bonita, passengers saw an eighty-foot-long black beast with white spots arc through the water.

Are Cryptozoological Animals-Real or Imaginary?

In local folklore these sightings have cumulatively been attributed to the mysterious existence of the Casco Bay sea serpent, an elusive monster ever roaming the cold Atlantic waters off the coast of southern Maine.

Beasts of the ocean have been lurking in human myth and legend for nigh on millennia, and their existence in the literary canon begins with some of our most ancient recorded stories. In the Rig Veda, a Hindu text composed around 1,500 B.C., Vritra is a water-hoarding dragon, defeated by the god Indra. Jörmungandr, the Midgard serpent, is the sea beast of Norse mythology, the archenemy of Thor. In the Babylonian creation myth, Enuma Elish, Marduk, a young god, defeats the dragon Tiamat, goddess of the seas and the embodiment of chaos. When she opens her jaws to devour him, Marduk causes a tempest to enter her mouth, and "the fierce winds fill her belly." With a single arrow, he then splits her in two. The Tigris and Euphrates Rivers flow from her eyes; from her body, he creates heaven and earth.

In Homer's Odyssey, Odysseus must navigate his ship between Charybdis, the treacherous whirlpool, and the terrible Scylla: "Twelve foul feet bear about her ugly bulk. Six huge long necks look out." She devours one of his men for each of her six heads. The Semitic deity Yamm, meaning "sea" in Hebrew, was ruler of the oceans and waterways. This same deity appears in the ancient Canaanite stories of Baal, myths etched into stone tablets, discovered in the ancient city of Ugarit in modern-day Syria. In this cosmogony, Yamm, the primordial chaotic force, is the god of storms and raging seas, slain by the hero Baal.

From Yamm we later get Leviathan, who makes a number of appearances in the Old Testament. In the Book of Job, a lengthy description of this untamable beast is used as a metaphor for the creative power and might of God:

> *It makes the depths churn like a boiling caldron*
> *and stirs up the sea like a pot of ointment.*
> *It leaves a glistening wake behind it;*
> *one would think the deep had white hair.*
> *Nothing on earth is its equal—*
> *a creature without fear.*
> *It looks down on all that are haughty;*
> *it is king over all that are proud. (Job 41:31–34)*

Here are a few sea monster stories. Some of these are unconfirmed reports, but they are good examples of these type of tales:

Are Cryptozoological Animals-Real or Imaginary?

Off the coast of Ireland on July 30th, 1915, the German submarine U-28 torpedoed the British ship Iberian. It went down rapidly, stern first. As the crew of the U-28 watched there was a large explosion that sent water and wreckage a hundred feet into the air. A "gigantic sea animal" was thrown to the surface and remained visible for about fifteen seconds before it sank. It was shaped like a sixty foot long crocodile with webbed feet.

Excerpt from the log of the ship General Coole, around 1780:

"A very large snake passed the ship. It was 3 or 4 feet in circumference. The back was of light color and the belly yellow." - S.H. Saxby, Master Mariner, Bouchurch, Isle of Wright.

In 1808 an Australian three-masted bark was attacked by a sea monster that, "had climbed across bow and bitten or chewed, one of the hands." It's eyes were the size of a "warrior's shield." The attack continued until the captain went below and returned with guns. He fired them into the animal's eyes and the monster returned to the ocean.

On December 7th, 1905 at about 10:15 am the oceanographic research yacht, Valhalla, was cruising off the coast of Florida and a "large fin, or frill, sticking out of the water," was spotted. The frill was six feet in length and projected almost two feet out of the water. "A great neck rose out of the water in front of the frill," noted Mr. Meade-Waldo, a scientist on board. The neck appeared to be about the thickness of a man's body. The creature moved its head and neck from side to side in a peculiar manner.

Three days after the Valhalla incident the Happy Warrior, a merchant sailing ship, reported a "sea snake of great

Are Cryptozoological Animals-Real or Imaginary?

magnitude appeared off our port bow. Was several lengths of our ship. Had long neck. Sounded after few minutes. Estimated speed six knots." The Happy Warrior was cruising only 80 miles from where the Valhalla sighted its creature.

A sea serpent, 45 feet long and 15 inches in diameter, was reported off the coast of Maine by Captain George Little in 1780:

"I was lying in Round Pond, in Broad Bay, in a public armed ship. At sunrise, I discovered a large serpent, or sea monster, coming down the bay. It was on the surface of the water. The cutter was manned and armed. I went myself in the boat. We proceeded after the serpent. When within a hundred feet, the mariners were ordered to fire on him. Before they could make ready, the serpent dove."

A U.S. Navy nuclear submarine left its home port to start its patrol. Mysteriously the boat's delicate sonar mechanism failed without warning only a few days into the voyage. The sonar was so critical to the sub's operations that the boat was forced to return to port for repairs. Examination of the sonar revealed that the rubber-like outer cover of the device had been torn off. Embedded in the tattered remains were enormous hooks. Scientists determined that these hooks, several times larger than had ever been seen before, were from a giant squid that had apparently attacked the sub, thinking it was a whale.

Toward the end of World War I the German submarine UB-85 was caught on the surface, during the day, and sunk by a British patrol boat. The crew abandoned the sub and was picked up by the British. The U-boat commander, Captain Krech was questioned about why he had been cruising on the surface and he told this tale:

Are Cryptozoological Animals-Real or Imaginary?

The sub had been recharging batteries at night on the surface when without any kind of warning a "strange beast" began to climb aboard from the sea. "This beast had large eyes, set in a horny sort of skull. It had a small head, but with teeth that could be seen glistening in the moonlight." The animal was so large that it forced the U-boat to list greatly to starboard. The captain feared an open hatch would drop below the waterline, flooding the sub and sinking it.

"Every man on watch began firing a sidearm at the beast," Krech continued. The animal had hold of the forward gun mount and would not let go.

The battle continued until the animal dropped back into the sea. In the struggle, though, the forward deck plating had been damaged and the sub could no longer submerge. "That is why you were able to catch us on the surface," the Captain concluded.

Evaluation of the Evidence:

There seem to be different types of sea serpents reported over the centuries. From giant squids to some types of giant snakes. The Kraken which seems to be some type of giant squid has also been mentioned in history.

We know very little about what animals live in the Oceans since we have barely explored below the surface and the world Ocean realm is many times the size of what lives on the land.

Are Cryptozoological Animals-Real or Imaginary?

It is easy for me to imagine that many huge unknown animals live in the Ocean. The largest animals we know like the Blue Whale live in our Oceans currently.

So it is certainly feasible for huge "Sea Monsters" to still be living in the Oceans. I would like to see one captured to give more information on what one looks like and its real size.

Are Cryptozoological Animals-Real or Imaginary?

13.0 The Chupacabra

The chupacabra is a legendary creature in the folklore of parts of the Americas, with its first purported sightings reported in Puerto Rico. The name comes from the animal's reported habit of attacking and drinking the blood of livestock, including goats.

Physical descriptions of the creature vary. It is purportedly a heavy creature, the size of a small bear, with a row of spines reaching from the neck to the base of the tail.

Eyewitness sightings have been claimed in Puerto Rico, and have since been reported as far north as Maine, and as far south as Chile, and even being spotted outside the Americas in countries like Russia and the Philippines, but many of the reports have been disregarded as uncorroborated or lacking evidence. Sightings in northern Mexico and the southern United States have been verified as canids afflicted by mange. According to biologists and wildlife management officials, the chupacabra is an urban legend.

Some recent sightings:

February 2017 ~ Victoria, Texas

On Highway 185 and Guadalupe Road in Victoria, Texas, a resident reportedly spotted a chupacabra. Following the tip of a Crossroads Today viewer, the news station went to the location and found an animal lying on the side of the road. It had "the paws of a dog, but the body of a hyena."

March 2017 ~ Olancho, Honduras

Residents of Las Agujas in Olancho, Honduras worried after finding a dead bull without eyes or a tongue. According to La Tribuna, residents saw an ugly white animal prowling the pastures believed to be El Chupacabra.

May 2017 ~ Choloma, Honduras

By May 6, 2017, a creature had killed 35 animals in the Monterrey de Choloma, a municipality in Cortés, Honduras. Residents feared that the animal – which disappeared as if by magic – would eventually hurt humans, especially children. Nely David Martínez saw it one day at 12:45 a.m. after hearing a noise. Struck by fear, he was unable to move and couldn't get a good glimpse at the figure, but all the animals began to drop to the floor, according to La Tribuna.

Are Cryptozoological Animals-Real or Imaginary?

May 2017 ~ Córdoba, Argentina

For months, people living in Charbonier in Córdoba saw animals attacked. In May, a man took a photo of an animal that he described as a "big bat, like the size of an eagle [that can] attack horses and cows," according to TKM. Specialists of the Servicio Nacional de Sanidad y Calidad Agroalimentaria (Senasa) dismissed Chupacabra claims and instead said it was bats that were transmitting rabies to other animals.

June 2017 ~ Nanegalito, Ecuador

59-year-old Casimiro Flores believes the creature he fought off was the Chupacabra. One day he heard a loud noise that sent chills up his spine. When he turned around, the about 5'7" man saw a creature – which looked like a brown dog with pointed ears – that reached his waist. Trapped by the animal, which dragged him around, Casimiro grabbed a stone and threw it at the creature's forehead. After letting out a cry, it ran away, according to Extra.

July 2017 ~ Riverside, California

In mid-July when Cary Shuker's cat raced inside their home, he looked outside and saw the "ugliest looking thing" staring at him about 80 feet away. With teeth jutting out in every direction, rippled skin, and a tail like a rat or possum, Shuker said it looked nothing like a coyote. It was also "at least two feet or more longer than the biggest coyote you've ever seen." Shuker's not the only one who has spotted the creature.

In the Box Springs Mountain territory.

Are Cryptozoological Animals-Real or Imaginary?

M.J. Bunt, an early childhood educator, also saw the chupacabra this year. "I thought, 'That is the strangest looking animal I've ever seen," she told The Press-Enterprise about spotting it near her home. "[It had] the ears of a deer, long snout, no hair, tail like a rat, long hindquarters. I thought it might be a sick coyote, a sick wolf. But it had too many different characteristics from any of them."

August 2017 ~ Santee, South Carolina

While golfing one day at the Santee Cooper Country Club, Doug Stewart took a photo of an animal that some believed to be a chupacabra. In a Facebook post that went viral, Stewart said the animal was most definitely not a dog. But as commenters weighed in with their different theories, some believed that it was actually a coyote or fox with mange.

Evaluation of the Evidence:

I have no doubt that people have seen something which corresponds to the description of the Chupacabra. My curiosity is about whether this is really a new animal or just a dog or wolf with serious mange. I've seen pictures of dogs with serious mange and the images I've seen of Chupacabras look very similar.

So, in this case I believe that there is no real species of Chupcabras that exists. That all sightings are of animals with serious cases of mange.

Are Cryptozoological Animals-Real or Imaginary?

14.0 Carnivorous Trees and Plants

All plants, unlike animals, are capable of producing their own food. They take carbon dioxide out of the air, water from the ground, light from the sun and make food through a process called photosynthesis. In addition to sunlight, carbon dioxide and water, the plants also need certain minerals to survive. These they usually take out of the soil though their roots.

Plants living in wet areas, such as bogs, have a problem, though. The water in these areas carry away many of the nutrients the plants need to grow. Some plants have found a solution to this problem by becoming carnivorous, which means "meat-eating." Rather than get the minerals they need from the soil, they trap animals, mostly insects, and take the nutrients out of the unfortunate victim's body.

Darwin originally called these flora insectivores. Later scientists decided that the plants ate enough animals (other than insects) that the term carnivore should be used.

Are Cryptozoological Animals-Real or Imaginary?

Meat-eating plants use several different mechanisms to trap their prey. Pitcher plants have leaves that grow into a vase-like container with a hood that overhangs the opening. The edge of the hood is covered with a sweet-smelling nectar that attracts insects. Inside this lid are downward-pointing hairs that lead the insects further into the plant and slippery wax surfaces that are difficult for the victims to crawl on. At the bottom of the vase-like structure is a pool of chemicals that will digest insects if they fall in. Various versions of the plants use different methods to get the victims into their pools. For example, the yellow trumpet has a substance in its nectar that paralyzes any insect that eats it. Once the victim takes a sip, he soon tumbles into the pool and is digested.

Other plants, like the sundew, use nectar to get insects to land on a leaf covered with sensitive hairs. Each hair has a tiny bead of sticky liquid on top. When the insect lands, it sticks to the hairs it is touching. As the victim struggles, other hairs bend over and attach themselves to secure the meal even further. The sticky liquid soon enters the insect's breathing holes and it suffocates. Digestive juices soon follow and the sticky liquid and the soft parts of the victim's body are soon dissolved away to be recovered and used by the plant.

Perhaps the strangest and most well-known carnivorous plant is the Venus flytrap. The flytrap, whose scientific name is Dionaea muscipula ("mousetrap of Venus") looks like a small circle of strange leaves sitting close to the ground. Sometimes it is topped by a long stem with small white flowers. These plants are so strange that folklore has it that they come from outer space and only grow near the sites of meteor impact craters.

Are Cryptozoological Animals-Real or Imaginary?

The truth is that the flytrap comes not from the second planet from the sun, but from North and South Carolina. It's strange leaves have a lobe at the end that looks like a small, green clamshell with teeth. Inside the clam shell are two sensitive hairs. If an insect lands on the lobe and touches both of the hairs, or touches one of the hairs twice in a short amount of time, the trap is sprung. The two sides of the clamshell leaf close quickly on the insect. The "teeth" intermesh, making sure the animal cannot escape. After the trap closes, glands on the inner surfaces of the shell release digestive juices.

Man-Eating Plants

Are any of these carnivorous plants capable of posing a threat to humans? Not really. The largest of the meat-eating plants is a relative of the pitcher plant named Nepenthes. It grows in the rain forests of Southeast Asia as a vine up to 50 feet in length. The pitchers sometimes grow to be a foot in length. Nepenthes traps mostly insects and small frogs, though animals as large as a rat have been found dead digesting in its juices. Some Nepenthes pitchers that have been found are large enough to hold four quarts of liquid.

The Nepenthes is not a threat to humans, however. In fact the local people have found ways to make them useful. The pitchers can be cleaned out and used to cook rice, while its long, strong vines serve as ropes.

If no carnivorous plant known is large enough to consume a human, where did the idea of man-eating plants come from?

Man-eating plants of South America', a carnivorous plant, can be found in the jungles of America mainly in the Brazil's Mato Grosso regions. The leaves of the plant hides beneath the forest floor and whenever a victim passes through it, they squeeze their prey till it dies. Similar kind of plant is found in those regions that have sweet berries like fruits in them to attract prey towards the plant. When the victim like birds and small animal's advances towards the tree, the branches closes and crushes the victim after which the blood is sucked by the tree's trunk.

The 'Cow Eating Tree' of India is also known as Padrame. In the year 2007, a cow was reportedly attacked by a carnivorous plant and the incident was encountered by a villager named Anand Gowda. The cow was seized by the branches of the tree and pulled it down to the ground. The local people also envisaged the entire scenario and some of them attacked the tree with knife and their axes. The local tribe of the village calls this tree as 'tiger tree' or 'Pili mara'.

Are Cryptozoological Animals-Real or Imaginary?

The Corpse Flower

The plant responsible for starting these rumors might be Amorphophallus titanum otherwise known as the "corpse flower." Amorphophallus titanum, which is said to be the biggest, smelliest flower in the world, looks like something that could eat a human being. When it blooms it can reach at height over nine feet in height and smells like a mixture of rotting flesh and excrement. The pungent odor attracts bees which are trapped in the flower until they are covered with pollen. Then they are released to fertilize other plants.

A blooming Amorphophallus titanum's "flower" (actually it is technically a leaf or spathe) can be three feet across. It is notoriously difficult to get a titanum to bloom outside of its native Indonesia, and botanical gardens around the world often try for decades without success. Bloomings of the Amorphophallus titanum have happened only about a dozen times in the United States since the first success at the New York Botanical Gardens in 1937.

Are Cryptozoological Animals-Real or Imaginary?

When the plant does bloom it moves quickly. It can grow as fast as 4 inches per day. The period when the "flower" is open lasts only about two days.

Although the Amorphophallus titanum looks a lot like you would imagine a man-eating plant to look like, and it even smells like somebody is dead inside, it is not carnivorous. Ironically, people are a lot more dangerous to carnivorous plants than the plants are to people. The pitcher plant and the venus flytrap live in wetland areas which are being destroyed by human development.

Also the popularity of these plants is also working against them. Commercial plant collectors have stripped areas of the Venus flytrap in an effort to provide them to consumers. Also many species of North American pitchers, whose leaves are in demand by florists, have been so decimated by collectors that they may disappear from the wild completely.

Evaluation of the Evidence:

Nowhere did I find a credible story of a plant eating a man. Yes-some plants do eat small animals and many eat insects, but to eat a person? No

The ideas of man eating plants are just not credible claims.

Are Cryptozoological Animals-Real or Imaginary?

15.0 Summary

Our world is filled with many strange and rare animals and there are still many places on the Earth or in the Oceans which can hide them.

Sea serpents could certainly hide in the Ocean which we have never fully explored ninety nine percent of it. And in spite of what people think there are many isolated areas on land where people rarely if ever go.

Studying these rare and fantastical animals and plants has given me a new perspective on what are real animals and what are imaginary.

I admit that I'm more open minded than many people about cryptozoology, but I do have my evaluation criteria. There are many sightings of the animals which I support, but there are questions about where they live and hide. These are good questions.

Let's keep making these observations of cryptids and looking for more evidence. It is searches like these that keep life exciting.

Martin K. Ettington
November 2019

Are Cryptozoological Animals-Real or Imaginary?

16.0 Bibliography

1. List of Cryptids. *https://en.wikipedia.org/wiki/List_of_cryptids.* [Online]

2. The Coelacanth Ancient Fish. *https://en.wikipedia.org/wiki/Coelacanth.* [Online]

3. Lists of Legendary Creatures. *https://en.wikipedia.org/wiki/Lists_of_legendary_creatures.* [Online]

4. Modern Peterosuars. *https://www.modernpterosaur.com/?cat=37.* [Online]

5. USA Pterosaur Sightings. *https://www.livepterosaur.com/LP_Blog/archives/category/usa-sightings.* [Online]

6. Giant Snake-Borneo River. *https://www.telegraph.co.uk/news/worldnews/asia/malaysia/4701906/Photograph-shows-giant-snake-lurking-in-Borneo-river.html.* [Online]

7. The Mapinguari. *https://en.wikipedia.org/wiki/Mapinguari.* [Online]

8. Giant Squid. *https://ocean.si.edu/ocean-life/invertebrates/giant-squid.* [Online]

9. Great Sea Serpent. *https://emergencemagazine.org/story/great-sea-serpent/.* [Online]

10. Sea Monster Stories. *http://www.unmuseum.org/tales.htm.* [Online]

11. Bigfoot Sightings Map. *https://www.nbcnews.com/sciencemain/looking-bigfoot-follow-map-others-have-seen-em-there-4B11203811.* [Online]

12. Most Convincing Bigfoot Sightings. *https://www.outsideonline.com/2097161/10-most-convincing-bigfoot-sightings.* [Online]

13. Mongolian Death Worm. *https://en.wikipedia.org/wiki/Mongolian_death_worm.* [Online]

14. The Cryptid Wiki. *https://cryptidz.fandom.com/wiki/Cryptid_Wiki.* [Online]

15. Werewolves from History. *https://www.cnet.com/news/wolves-among-us-five-real-life-werewolves-from-history/.* [Online]

16. Man Eating Tree. *https://en.wikipedia.org/wiki/Man-eating_tree.* [Online]

17. Man Eating Plants. *http://www.unmuseum.org/maneatp.htm.* [Online]

18. American Indian Thunderbird Legends. *http://www.native-languages.org/thunderbird.htm.* [Online]

19. Unicorns. *https://mysteriousuniverse.org/2018/03/strange-encounters-with-unicorns/.* [Online]

20. The Chupacabra. *https://en.wikipedia.org/wiki/Chupacabra.* [Online]

21. Real Dragons. *https://mysteriousuniverse.org/2019/02/a-strange-history-of-real-dragons/.* [Online]

22. The Kraken. *https://en.wikipedia.org/wiki/Kraken.* [Online]

Are Cryptozoological Animals-Real or Imaginary?

Are Cryptozoological Animals-Real or Imaginary?

17.0 Index

Bigfoot, 3
Carnivorous Trees, 101
Chupacabra, 97
Cryptozoology, 1
Dinosaurs, 53
Dragons, 29
Giant Snakes, 47
Giant Squids, 88
Man-Eating Plants, 103
Mapinguari, 83
Mokele-Mbembe, 53
Mongolian Death Worms, 59
Native American Thunderbird, 18
Pterodactyls, 13
Real Dragons, 29
Sasquatch, 3
Sea Serpents, 87
The Bunyip, 25
The Corpse Flower, 105
The Livonian Werewolf, 75
The Werewolf of Allariz, 78
The Werewolf of Bedburg, 80
The Wolf of Ansbach, 77
Thunderbirds, 13
Unicorns, 63
Werewolves, 73

Are Cryptozoological Animals-Real or Imaginary?

www.ingramcontent.com/pod-product-compliance
Lightning Source LLC
Chambersburg PA
CBHW030702220526
45463CB00005B/1871